WELCOME TO THE UNIVERSE

A POCKET GUIDE FOR VISITORS

KEVIN J SIMINGTON

ISBN: 978-0-6484945-4-6 (Color Edition)

❀ Created with Vellum

CONTENTS

1

THE SOLAR SYSTEM IS BIG

So, you've been born into the universe, have you? Congratulations! You're already a big step ahead of the quintillions of people who've never been born at all. Those poor sods are just drifting around in oblivion doing nothing. That has to be even more boring than filling out your tax return.

You might not think that being born was a particularly impressive achievement, but you're failing to appreciate the monumental odds that you overcame just to be here. You came first out of about 100 million sperm! The statistical improbability of that is staggering, especially considering the fact that a single sperm is only ever the fastest and strongest for a few short seconds. With 1,500 new sperm being produced every second inside a man's testes, faster stronger new sperm are constantly superseding the old sperm which are gradually losing mobility and dying. You came into existence at exactly the right time! And your triumphant moment coincided perfectly with your parents' triumphant moment. If your parents hadn't had their glorious moment exactly when they did (sorry to bring this up – are you eating dinner?) - if it had been just a few minutes earlier or a few minutes later (for example if your Aunt Mildred had interrupted their warm up routine by phoning to ask for your great-grandmother's Christmas cake recipe) - you would not have been at your peak performance as a virile young sperm. A few moments either side of that momentous copulative event, and you would have been pipped at the post by a stronger, faster sperm, and you would never have had to fill out a single tax return.

Anyway, now that you're here, I suppose I should tell you a bit about the universe. And the sooner the better – because you're not going to be around for very long, are you? But let's not get into that.

The universe is big. I mean, *really* big. Gob-smackingly BIG! If you could drive from Earth to the edge of the known universe in your family car, you would definitely need to pack more than a picnic lunch.

In order to explain how big the universe is, I really should start in our own backyard – our solar system. A solar system is a system of planets revolving around a star (also called a sun – hence the word "solar").

Our solar system has either 8 or 9 planets, depending on which scientist you're speaking to and what sort of mood they're in. Pluto was demoted from the 9th planet to a 'dwarf planet' a few years ago, which really ticked off all the inhabitants there, I can tell you! So, taking into account Pluto's sad demise, our solar system consists of:

- 1 Star (the Sun)

- 8 planets (Mercury, Venus, Earth, Mars, Jupiter, Saturn, Uranus and Neptune)
- 5 dwarf planets (Pluto, Ceres, Haumea, Makemake and Eris)
- 181 moons (Jupiter has 79, Saturn – 62 and Uranus – 27)
- 566,000 asteroids
- 3,100 comets

(Image: the planets.org)

You might be tempted to think that with all these celestial bodies whizzing around, the solar system must be a pretty crowded place. Not so. There's a heck of a lot of space between all of these bodies.

For instance, the distance from the Sun to the Earth, the third planet in the solar system, is 150 million kilometres. That's 8 light-minutes away. What does that mean? Let me explain. Light travels at about 300,000 kilometres per second. That's right, per SECOND. That's eight times around the earth in a single second. But the Earth is so far from the Sun, that light emanating from the Sun, travelling at that astonishing velocity of 300,000 kilometres per second, takes 8 whole minutes to reach the Earth. This mean, by the way, that we are actually seeing the Sun as it was 8 minutes ago. When we look at the sun (hopefully not with the naked eye) we are actually looking into the past, because we are seeing the Sun as it was 8 minutes ago. The Sun could blow up right now, and we wouldn't know about it for 8 more minutes.

To give you another way of appreciating the distance of the Earth to the Sun, consider the speed of a bullet. The initial velocity of an average bullet is about 2,000 kilometres per hour. If we fired a bullet at the Sun, and its velocity never retarded, it would take over 8 YEARS to reach the Sun!

But the distance of the Earth to the Sun is a relatively unimpressive distance when compared to the size of the whole solar system. The very outer edge of our solar system, beyond the orbits of the comets where the Sun no longer exerts a gravitational force, is about 14.5 billion kilometres from the Sun. A bullet fired from Earth towards the solar system's outer edge would take 827 years to reach there! Light from the sun, travelling at 300,000 kilometres per second, takes 2 years to reach the solar system's outer edges.

These vast distances pose massive problems for space travel. Space travel to the planets within our own solar system will involve YEARS of space flight. The fastest manned spacecraft to date was Apollo 10, which reached a whisker under 40,000 kilometres per hour on its return journey from the moon, which is 20 times the speed of a bullet. That's fast! But even travelling at that impressive velocity, the following transit times would be required for travelling to the various planets in our solar system:

- Mars – 1 year 2 months
- Jupiter – 2 years 6 months
- Saturn – 4 years 1 month
- Uranus – 8 years 3 months
- Neptune – 12 years 10 months
- Outer edge of the solar system – 41 years

Does that stagger you? Travelling at the fastest velocity that scientists have so far achieved for a manned space flight, which is 20 times the speed of a bullet, it would take astronauts 41 YEARS to reach the edge of our solar system and make it into interstellar space – the space between the stars.

(Image courtesy of NASA: Apollo 10)

To give you another way of gauging the size of our solar system, we can consider the orbital periods of the planets. This is the time that it takes each planet to complete a single orbit around the Sun. The Earth takes 1 year to orbit the Sun. But the outer planets are much further from the sun, so it takes them much longer. The orbital periods of the 8 planets plus Pluto are shown below:

- Mercury – 87 Earth days
- Venus – 224 Earth days
- Earth – 1 year
- Mars – 1.9 Earth years
- Jupiter – 11.8 Earth years
- Saturn – 29.5 Earth years
- Uranus – 84 Earth years
- Neptune – 164.8 Earth years
- Pluto – 247.7 Earth years

It takes poor little Pluto over 247 Earth years to complete a single orbit of the sun. And that is simply because it is so far away.

The solar system is huge!

❧

2

A QUICK GUIDE TO THE PLANETS

The Earth is the third planet from the sun and is a pretty decent place to live, apart from difficulty finding parking spaces and those really annoying tear-off tabs on washing powder boxes that never tear off properly. It's also the only planet in our solar system with liquid water, a breathable atmosphere, a liveable climate and Netflix.

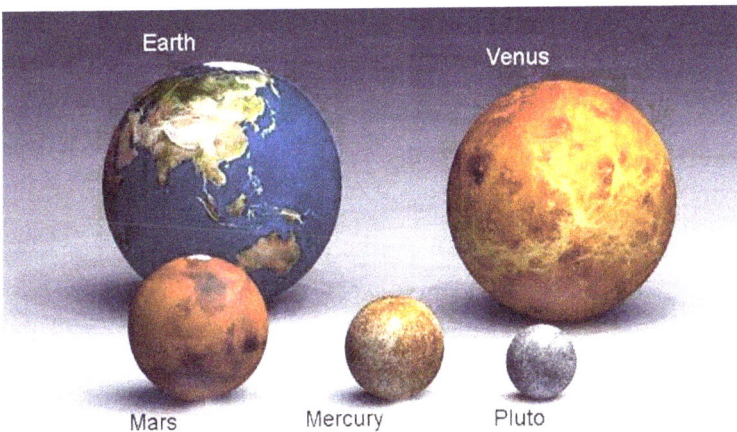

(Image: universe today.com)

The image on the previous page shows the comparative size of Earth and the smaller planets in our solar system. Note how small Mars is; less than half the size of Earth. Compared to the other inner planets, Earth is a pretty decent size.

But when we start comparing our planet to the larger celestial bodies in our solar system, we start to appreciate the sheer scale of the planets. The following image shows the comparative sizes of ALL the planets, plus Pluto.

(Image: universetoday.com)

Jupiter, the largest of the planets, is bigger than all the other planets combined. If all the other planets were melted together, they would still not be as big as Jupiter. If Jupiter was hollow you could fit 1,300 Earths inside it!

But Jupiter is not the biggest body in our solar system. That award goes to the Sun. The image below shows the size of the Sun when compared to the planets. At this scale, the Earth is just a tiny dot at the bottom of the image.

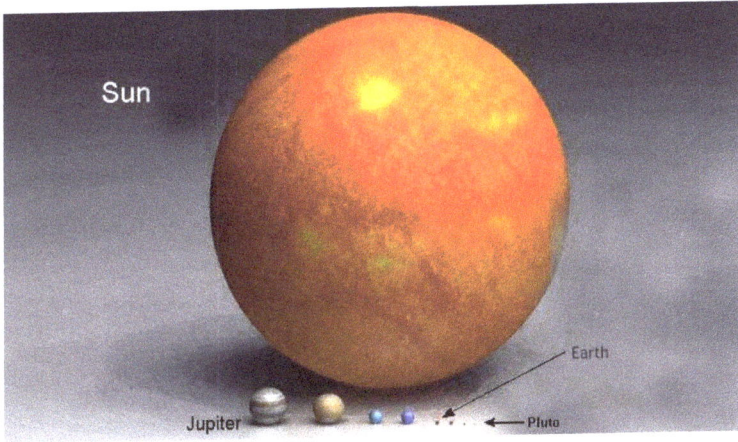

(Image: astroobserver.com)

The Sun is so massive, if it was hollow it could fit 1.3 million Earths inside it!

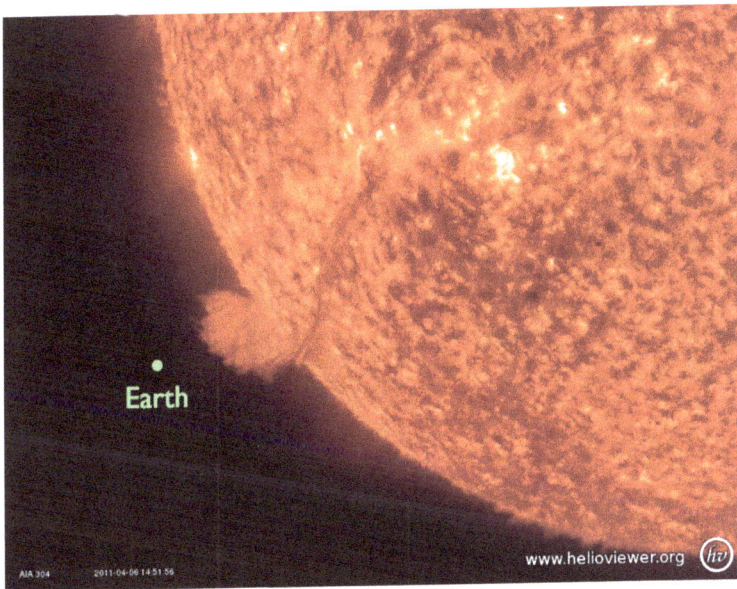

***

Here are some Interesting Facts About the Planets and Moons:

Mercury

- The smallest planet in the solar system, with a diameter of only 4,879 km.
- Daytime temperature is +430 Celsius (800 F) but drops to -180 C (-290 F) at night.
- A rocky planet with a thin atmosphere.

Venus

- The hottest planet in the solar system. Daytime temperature is +465 C (900 F), which is hot enough to melt lead.

- It rotates in the opposite direction to the other planets.
- Poisonous and corrosive gases in the atmosphere and hurricane force winds.
- Venus is the brightest object in our night sky, apart from the moon.
- It rotates VERY slowly on its own axis. One day on Venus takes 243 of our Earth days, whereas it orbits the sun in 225 of our Earth days. Thus, a day on Venus is longer than a Venus year.

Earth

- The only planet in the solar system with Netflix and Wi-Fi internet.
- Limited parking.
- Washing powder boxes difficult to open.
- 71% water (which would be really handy if we could work out how to open the washing powder boxes).
- Home to the only known intelligent life in the universe, which doesn't say much for the rest of the universe really.

- The only known planet with a breathable atmosphere

Mars

- The second smallest planet in the solar system.
- Half the diameter of Earth and 10% of Earth's mass.
- Gravity is only 37% of Earth's, so you could jump 3 times higher on Mars.
- Has the tallest mountain in the solar system – Olympus Mons, 21 km (13 miles) high (Everest is 8.8 km high).
- Has a thin atmosphere of carbon dioxide, nitrogen and argon, and atmospheric pressure on the surface is only about 1% of Earth's.
- The surface has channels that appear to have been carved out of it, either by wind or water.
- Temperatures range from -87 to -5 C.
- Has 2 moons.

Jupiter

- It is a gas giant, not a solid rocky planet.

- The largest planet in the solar system; greater in mass than all the other planets in the solar system put together.
- 1,300 Earths could fit inside it.
- Has at least 79 moons. Its biggest moon, Ganymede, is larger than Mercury.

- Has the shortest day in the solar system; rotating in just under 10 Earth hours.
- Has an enormous dark storm, 25,000 km in diameter – twice the size of Earth - that has been raging for over 300 years.
- Temperature is -108 C.
- Wind speeds up to 620 km per hour.
- Takes 11.8 Earth years to orbit the Sun.
- Its gravity is 2.36 times Earth's gravity.
- It is the fastest spinning planet, having a rotation speed of 43,000 km per hour at its equator (compared to Earth's 1,600 km per hour).

Saturn

- Gas giant, 95 times the mass of Earth.
- Fastest winds in the solar system; 1,800 km per hour.
- Temperature is -139 C.
- Rings consist of ice and rock; 250,000 km in diameter but only 1 km thick.
- The rings have radial 'spokes' which may be a result of Saturn's magnetic field.
- Has at least 62 moons.
- Its yellow colouring is due to Ammonia crystals in its upper atmosphere.
- It takes 29.5 Earth years to orbit the Sun.
- It has the second fastest rotation speed of 35,000 km per hour at its equator, (compared to Earth's 1,600 km per hour).
- It is the furthest planet that can be viewed with the naked eye.

Uranus

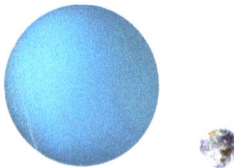

Comparative sizes of Uranus and Earth

- An ice giant, 14 times the mass of Earth
- The only planet that rotates perpendicular to the orbital plane of the solar system.
- Takes 84 Earth years to orbit the Sun.
- Temperature is a cool -189 C.
- The predominance of methane gives it a blue colouring.
- Has 27 moons.

Neptune

- A gas giant, 17 times the mass of Earth.
- Temperature is a cold -214 C.
- Has 14 moons.
- Takes 164.8 Earth years to orbit the sun.
- Has EXTREMELY high wind on its surface – about 1,340 km per <u>second</u>!
- Is coloured blue because of its methane atmosphere.

Pluto

- Even though it's no longer considered to be a planet, it's still a very cool place – literally; -280 Celsius.
- Like Venus, it rotates in the opposite direct to the other planets.
- It is only 2,373 km in diameter, which is smaller than our Moon.
- It takes 247.7 Earth years to orbit the Sun.
- It only takes sunlight 8 minutes to reach Earth but 5 hours to reach Pluto.

Io

- A moon of Jupiter, slightly larger than Earth's moon.
- Io is the most volcanically active body in the solar system.
- Hundreds of volcanoes and volcanic vents are regularly erupting magma.

- Some eruptions are so explosive that magma is ejected from the planet and ends up on Jupiter.
- Io's volcanic activity arises from the strong gravitational forces from Jupiter and several large moons which are constantly distorting Io and creating friction within the mantle and core of the moon, grinding and melting rock.
- It is believed that over its life, Io has erupted more than its own mass (as it recycles magma that cools again).

Titan

- The second largest moon of Saturn and the only known moon with an atmosphere.
- 1.5 times the size of Earth's moon.
- Its atmosphere is denser that Earth's and is comprised of 90% nitrogen and 10% a mixture of other gases including methane and ethane.
- Larger than the planet Mercury but only 40% of Mercury's mass.
- Temperature is -180 C.
- Contains vast amounts of water ice (H_2O).

- Probably has a core that is still hot or possibly even molten.

Europa

- Moon of Jupiter.
- It is slightly smaller than Earth's moon and has the smoothest surface of any celestial body in the solar system.
- It is believed to have a liquid water ocean under the surface.

- Tidal and gravitational forces keep the water in a liquid form.
- Plumes of water and water vapour, called cryogeysers, have been seen erupting from the surface and sending plumes high into the atmosphere.

3

COLONISING OTHER PLANETS

It is almost inevitable that mankind will eventually establish permanent bases or colonies on other planets within our solar system. Apart from the scientific value of doing this, there is also the very practical motivation of mining other planets for valuable metals and minerals. The two most likely candidates for human colonisation are the Moon and Mars.

The moon is a rich resource for mining, containing large deposits of silica (SiO_2), alumina (AlO_3), lime (CaO), iron oxide (FeO), magnesia (MgO), and titanium dioxide (TiO_2). This means that the most abundant element on the moon is oxygen, locked in these oxygen-bonded minerals and metals. Because of this, ample breathable oxygen could be produced as a natural by-product of the mining and manufacturing processes.

The moon also has several areas that contain subterranean hollow lava tubes which could be utilised and adapted for human habitation. This would provide protection from the dangerous solar radiation that bombards the surface of the Moon because it doesn't have a protective magnetic field.

(Image: ESA.int)

Water ice has also been detected on the moon; on the surface at both poles. There is also speculation as to whether a region of the moon may contain subterranean ice. The downside of a colony on the Moon would be the low gravity, which would necessitate colonists having regular gym sessions to simulate Earth gravity. Its proximity to Earth, however, makes it a prime target for a colony. Nuclear fusion plants could provide power on the Moon, using the helium-3 isotope which is in abundance.

In my science fiction novel, "The Stars That Beckon" the first in the StarPath Trilogy, I incorporated a Moon base (Armstrong Base) built into the Moon's lava tubes and utilising the vast bed of water ice under the surface.

(Image: nst.com.my)

Mars is the other obvious choice for a human colony. Although it is further away, requiring over a year of space travel, it has the advantage of a higher gravity. It has deposits of ice at its poles and also huge reservoirs of subterranean ice – an estimated 21 million km^3. Mars has an abundance of minerals such iron, titanium, nickel, aluminium, sulphur, chlorine and calcium, which make it a prime candidate for a mining base. Silicon oxide is also present in abundance, which would make it possible to produce glass and fiberglass for construction of habitats and structures. Habitats would need to be buried underground or located in caves or deep shafts, in order to protect the colonists from deadly solar radiation.

The limiting factor in the colonisation of either Mars or the Moon, is our ability to devise an efficient and cheap propulsion system for space travel. To make space travel viable, a propulsion system needs to be developed that can accelerate a spacecraft continuously for the first half of the journey and decelerate it continuously for the second half. This would dramatically cut down transit times. Rockets and spacecraft will also need to be completely reusable in order to cut down costs. Once such a system is developed, we could see viable off-world bases established very quickly.

(Image: straitstimes.com)

Over the last 20 years, a number of Mars missions have been proposed. Mars One, proposed in 2012, arguably received the most publicity. It proposed sending robots to Mars and establishing basic infrastructure ahead of the first human colonists who were scheduled to arrive in 2025. A short list of crew was even selected out of thousands who applied, on the understanding that it would be a one-way mission. However, the Mars One company went bankrupt in January 2019.

Currently NASA, SpaceX, ESA (European Space Agency) as well as various national space agencies such as India, Japan, China and Russia all have plans to eventually establish a human base on Mars. SpaceX, headed by Elon Musk, currently plans to send unmanned cargo vessels to Mars in 2022 to set up basic infrastructure and then send a human crew in 2024.

In my science fiction novels, "The Stars That Beckon" and "A Path Through The Stars", the first and third books in the

StarPath Trilogy, I incorporated a Mars base built at the bottom of a large crater which provided shelter from cosmic radiation.

The challenges involved in colonising planets in other solar systems is exponentially greater than the challenges we face within our own solar system. This is because of the inconceivably vast distances involved. We will look at this problem in a later chapter.

$\backsim$

4

A QUICK GUIDE TO THE STARS

Stars or suns are bright balls of gas, mostly hydrogen and helium, that are so large and dense that an enormous nuclear fusion reaction is taking place at their core, producing vast amounts of heat, light and radiation. Our Sun is the largest body in our solar system by far. If you clumped together all the planets, all the asteroids, all the moons and all the comets in our solar system, they would amount to only 0.14% of the solar system's mass. The remaining 99.86% is the Sun!

This is why all of the other celestial bodies in our solar system revolve around the Sun. The Sun's enormous mass exerts a huge influence on the planets, asteroids comets and moons, trapping them all in an endless orbit around itself.

In terms of its size, the diameter of the sun is approximately 1.4 million kilometres, which is about 110 times the diameter of the Earth. Roughly 1.3 million Earths could fit inside the Sun!

(Image: ppl.nasa.gov)

Temperatures inside the sun can reach 15 million degrees Celsius. The inner two inner layers of the Sun's atmosphere – the photosphere and chromosphere - are much cooler, only about 5,000 and 4,000 degrees respectively.

Surprisingly, the Sun's third and outer layer of atmosphere, the corona, is its hottest part – 17 million degrees Celsius. This is hotter than the sun's core. Scientists have several theories why this is so, but they are still not completely sure.

Huge explosions on the Sun's surface and extending up through its atmospheric layers cause solar flares. These send immense burst of deadly radiation flooding through the solar system. This is on top of the already constant flood of radiation being emitted through the Sun's ongoing nuclear reaction. Earth's magnetic field protects us from the worst of this radiation, deflecting this 'solar wind' around our planet.

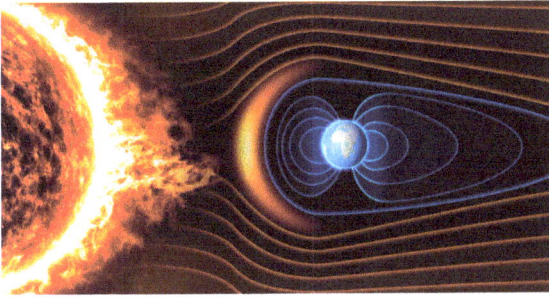

(Image: theburnin.com)

The famous northern lights (Aurora Borealis) and southern lights (Aurora Australis), visible in the night sky in our far northern and southern hemispheres, are the Sun's lethal radiation bombarding the Earth's magnetic field and being deflected harmlessly away into space.

(Image: photographylife.com)

In terms of the relative size of our Sun, it is dwarfed by much bigger stars in the universe. Our Sun is a yellow dwarf, but there are other types of stars that are much bigger. The following image shows the relative size of our sun when compared to the stars, Sirius, Pollux and Arcturus. Arcturus, a red giant in the Bootes Constellation, is 23,000 times our Sun's mass! At this scale Jupiter is about 1 pixel and Earth is invisible.

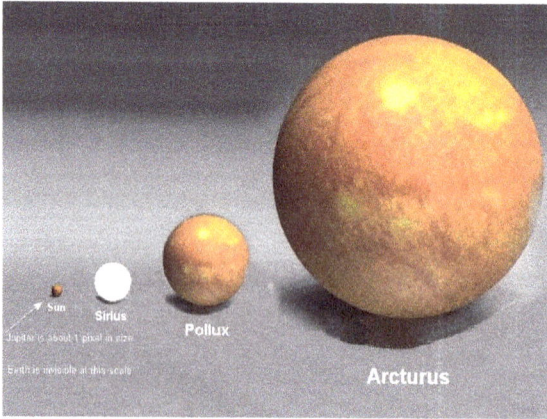

(Image: jpl.nasa.gov)

But there are much bigger stars than Arcturus. The following image shows the same stars, the Sun, Sirius, Pollux and Arcturus, compared to Betelgeuse and Antares, red super giants.

(Image: jpl.nasa.gov)

Antares could fit 630 million of our Suns inside it! At this scale our Sun is now only 1 pixel on the page – no longer visible.

But Antares isn't the biggest star we have found. The largest star discovered to date is the red hypergiant, VY Canis Majoris.

The name Canis Majoris is Latin for "big dog", and that's exactly what it is – the big dog of all stars! It has a volume 3 BILLION times our Sun and a mass 9.8 BILLION times our Sun! Canis Majoris is an oxygen-rich star that, because of its enormous size, is the most luminous star in the Milky Way.

The following image shows its enormous size when compared to Antares. At this scale, our Sun is too small to be represented - about one thousandth of a pixel!

(Image: jpl.nasa.gov)

To help you understand just how big Canis Majoris is, if we could transport it to our own solar system, it would encapsulate the entire solar system – all the way out to the orbit of our furthest planet! In fact, in terms of its three-dimensional volume, you could fit 2,000 of our solar systems inside it, stacked on top of each other! To put it another way, you could fit 7 quadrillion Earths inside it.

Now, 7 quadrillion is a number that is very difficult to grasp. So let me help you, by using time:

- 1 million seconds ago = 12 days ago
- 1 billion seconds ago = 33 years ago
- 1 trillion seconds ago = 29,000 BC
- 1 quadrillion seconds ago = 34 million BC (if such a time ever existed!)

A star big enough to fit 7 quadrillion Earths inside? Now that's BIG!

∽

5

GETTING AROUND THE GALAXY

O ur solar system is dwarfed by the size of our galaxy. A galaxy is a collection of billions of stars, often with planets orbiting them, like our own solar system. Our Milky Way Galaxy contains about 200 billion stars.

(Image: pinterest.com)

The image above shows a dot where our solar system is located within the Milky Way. However, if the picture was drawn to scale, our entire solar system, including the sun and all the planets, would be so small it would be invisible - less than a thousandth of a pixel!

The Milky Way Galaxy is HUGE. It is 100,000 lightyears in diameter and 1,000 lightyears thick. This means that light, travelling at 300,000 kms per second (yes, per SECOND, not per hour), takes 100,000 years to travel from one side to the other!

The stars in the Milky Way are not evenly distributed. They tend to cluster in spiral arms, and the whole galaxy is spinning around its galactic core. The image above looks like a swirling mass of cloud, but those 'clouds' are simply billions of stars. The galactic core is so densely populated with stars that no life would be possible on any planets there, due to the deadly radiation that permeates that region. If our solar system was located closer to the galactic core, we would not be able to survive; even the simplest of biological life would be wiped out by the vast amounts of deadly radiation. Our solar system, however, is located in a safe, less densely populated region, on the outer edge of one of the spiral arms.

Sideview of a galaxy. (Image:jpl.nasa.gov)

When we look up at the belt of stars across our night sky, we are looking at the side-view of our Milky Way Galaxy. In good viewing areas, without city lights, it is possible to see the expanse of the galaxy with the thicker galactic core in the centre.

Our galaxy, seen from Earth. (Image: ESA.org)

Scientists have recently discovered that our galaxy has a black hole at its centre, as, indeed, do most galaxies. A black hole is an object of such dense mass that its gravity won't even allow light to escape.

(Image: solarsystem.nasa.gov)

It is thought that black holes form when a star collapses upon itself. The black hole at the centre of the Milky Way Galaxy exerts a huge gravitational force on the entire galaxy, keeping the billions of stars within the galaxy orbiting around the centre.

The sheer size of our galaxy poses huge problems for potential interstellar space travel. The nearest star to our own Sun is Proxima Centauri, which is 4.22 lightyears distant. This means that light, travelling at 300,000 kms per second, takes 4.22 years to travel that distance. Our problem, of course, is that we aren't able to build spacecraft that can travel anywhere near the speed of light. The fastest manned spacecraft we have produced to date, Apollo 10, travelled at a velocity of a mere 40,000 km per hour (20 times the speed of a bullet). That is only 0.000037% of the speed of light. Travelling at that speed, it would take a spacecraft 113,940 years to reach Proxima Centauri. And that is the closest star to us! Travel times become astronomical (literally!) when we consider travelling to stars much further away. Even if we could build a spacecraft that could achieve 1% of the speed of light (that's 30,000 times faster than Apollo 10 and 600,000 times the speed of a bullet), it would still take 422 years to reach Proxima Centauri – and that is our nearest star. And travelling to the far side of our galaxy, even at that speed, would take 10 million years!

This is why some scientists declare that manned interstellar space flight is impossible. Others dream of the possibility of a crew being placed in some kind of cryogenic stasis or deep freeze where they could sleep the centuries away while their spacecraft sailed across the interstellar void, only waking when the ship reached its destination. I incorporated this concept in my science fiction book, "*The Stars That Beckon*" and the two following novels in the StarPath Trilogy, "*The Stars That Bend Time*" and "*A Path Through The Stars*". The possibility of such technology, however, remains highly speculative.

∾

6

THE NEVER-ENDING UNIVERSE

W e haven't finished exploring just how big our universe is. So far, I've only described the size of our local galaxy – the Milky Way. That's really just like describing the back garden of your house, compared to the whole world. You see, our galaxy is only one of billions of galaxies in the universe, each of them containing billions of stars.

There are all kinds of spectacular galaxies out there. Thanks to the Hubble Space Telescope, scientists have been able to capture extraordinary images of some of these galaxies over the last few decades.

The photographs on the following pages may look like clouds of gas, but they are actually galaxies, with billions stars creating an impression of a gas or dust cloud. The phrase, "stars like dust", popularised many decades ago by Isaac Asimov's science fiction novel of that name, is entirely appropriate.

The Andromeda Galaxy (Image: nasa.gov)

The Sombrero Galaxy (Image: nasa.gov)

A Lenticular Galaxy (Image: hubblesite.org)

Colliding Galaxies (Image: hubblesite.org)

Colliding Galaxies (Image: hubblesite.org)

*Interacting Galaxies (one galaxy stripping stars from
another) (Image: hubblesite.org)*

The Antennae Galaxies (Image: hubblesite.org)

More Interacting Galaxies (Image: hubblesite.org)

The Whirlpool Galaxy (Image: hubblesite.org)

The Ring Galaxy (Image: hubblesite.org)

Our closest galaxy, the Andromeda Galaxy, is 2.5 million lightyears away. That means that light takes 2.5 million years to travel between our two galaxies. Even if we could build a space-craft capable of travelling at the speed of light, it would take us 2.5 million years to reach Andromeda! And that is our nearest galaxy neighbour!

To give you an idea of just how big the universe is and how many galaxies there are out there, let me tell you a story. In 2009, the Hubble Space Telescope, orbiting around the earth, underwent a major refit.

(Image: nasa.gov)

NASA fitted it with a much higher resolution imaging system, enabling it to see further into space than ever before. Once the refit was complete (after a few technical hitches) NASA focussed the telescope on a tiny patch of space that appeared to have no stars in it; a tiny area of complete blackness.

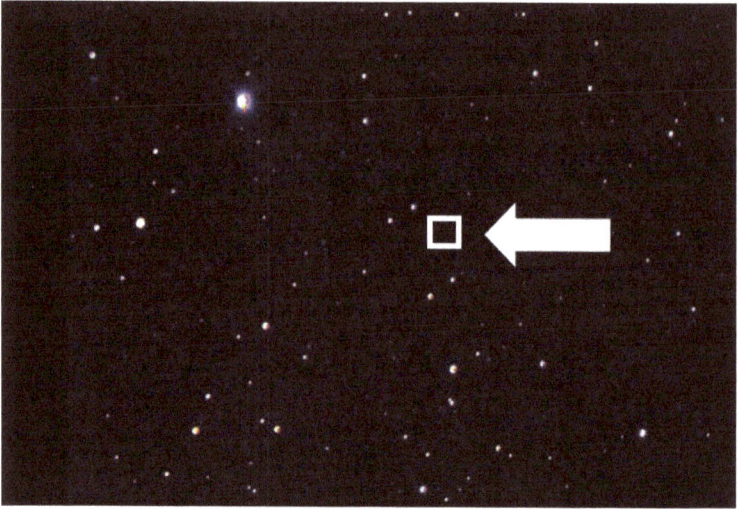

(Image: hubblesite.org)

Then they activated the new imaging system, and this is what the telescope revealed:

(Image: hubblesite.org)

That tiny area of blackness suddenly became filled with

galaxies! Yes; with galaxies, not just stars! That's right! Apart from about 6 dots of light in the photograph above, every point of light in that picture is an entire galaxy – each with billions of stars. And all of those galaxies are in a tiny portion of the night sky that would be smaller than the size of the fingernail on your little finger if you held your hand out at arm's length!

The Hubble Space Telescope has revealed to us the mind-boggling size of our universe. There are literally trillions of galaxies out there, each with billions of stars, and, as far as we can see, the universe has no limit. We have not yet been able to see even the hint of an outer edge to it.

So, next time you feel big and important, stand outside at night, and look up at the stars. It will really put you in your place!

~

7

CHALLENGES OF HUMAN SPACE FLIGHT

F uture human space flight to other planets faces 4 main challenges (plus a 5th if the 4th challenge can be overcome).

1. RADIATION

Radiation is an invisible but deadly challenge facing long space flights. Once humans leave the protective shielding of Earth's magnetic field, their exposure to radiation increases enormously. The Sun is constantly bombarding our universe with huge amounts of dangerous radiation, including deadly gamma radiation. The Sun's extreme radiation in space has the potential to cause cancer, damage the central nervous system and significantly impair cognitive and motor function. Astronauts within our international space station, which orbits just *inside* Earth's magnetic field, experience 10 times the radiation levels than people do on the surface of our planet. In space *beyond* our magnetic shield, the level of radiation will be

HUNDREDS of times more than on Earth. And during a solar flare, when a massive explosion occurs on the surface of the Sun, a wave of deadly radiation that is THOUSANDS of times more potent will sweep through our flimsy spacecraft.

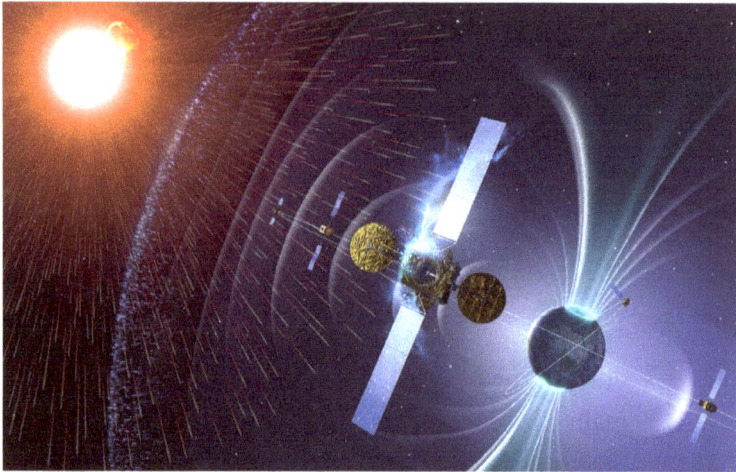

(Image: ESA.int)

In order to overcome this serious problem, three solutions are currently being investigated. The first involves significantly increasing the physical shielding of spacecraft. The problem with this, of course, is that it would also significantly increase the mass of the spacecraft, which in turn increases the amount of fuel needed for propulsion. And more fuel equals even more weight again. The second area of research involves medications that could possibly reduce the physiological impact of radiation on the human body, but such medications will almost certainly have side effects for those taking them. The third proposal is to build into each spacecraft a heavily lined 'safe room' where crew and colonists can temporarily take refuge when a solar flare occurs.

. . .

2. REDUCED GRAVITY

After many decades of studying the effects of low or no gravity on astronauts, some considerable problems have become apparent. Physiological effects of weightlessness on the human body include lowered bone density, osteoporosis, loss of muscle mass, loss of balance, digestive problems and various metabolic issues. The longer a person's body is exposed to weightlessness, the more pronounced these problems become. For example, if a spaceship was travelling for several years to Jupiter or Saturn with no gravity on board, by the time the crew arrived at their destination their bones would be extremely brittle and they would have lost 90% of their muscle mass.

(Image: nasa.gov)

There are basically three ways of counteracting the impacts of weightlessness. The first is simply to accelerate at 1G for the first half of the journey and decelerate at 1G for the second half. This is currently impossible, however, because of our inability to carry that much fuel on board a spaceship! The second possibility is to have a gym on board that has exercise machines that simulate gravity. The International Space Station has such a gym, using large elastic bands that are strapped around an astronaut pulling them down onto a treadmill as they run. This is only mildly effective, however, because there is only so much time in the day when a person can exercise, and even then, these machines don't exert simulated gravity on every part of the body.

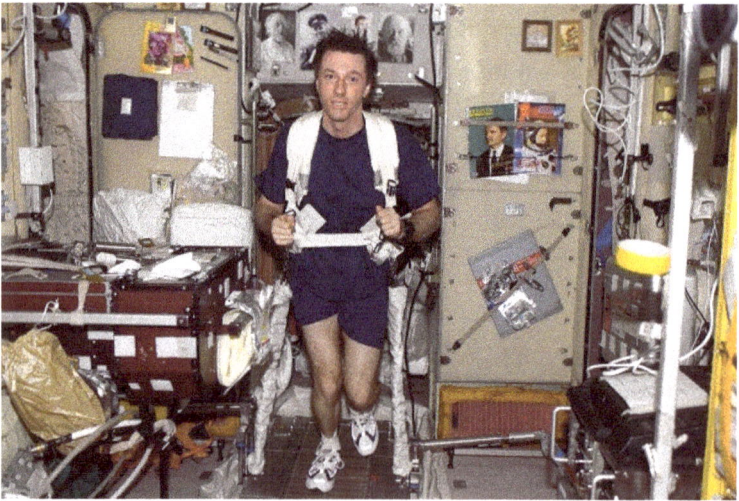

(Image: nasa.gov)

The third, and best, solution would be to build a rotating section of the spacecraft which would produce 1G gravity via centrifugal force, where crew and colonists could live for the duration of their journey. This, however, would involve a much more complicated and expensive type of spacecraft.

. . .

3. ISOLATION and CONFINEMENT

Psychological and social problems are almost inevitable when you cram groups of people together in a very confined space for long periods of time. On Earth, even in our most confined spaces, we can still make a phone call or connect with someone outside our confinement with various forms of technology. But in long-term space flight, that will not be possible. Huge time delays in communication, involving many hours, days, weeks, months or even years (depending on how far from Earth the spacecraft is) will make that kind of outside communication impossible.

(Image: nasa.gov)

Long-term isolation and confinement can result in a range of health problems including sleep loss, circadian desynchronization, depression, boredom, crew conflict and even such things as psychosis.

In order to minimise the risks of these problems, a crew will need to be highly trained and given strong coping mechanisms. Crews will also need to be very carefully chosen to minimise potential interpersonal conflict. Other strategies include entertainment (such as a library of games and films) and provision for regular physical exercise.

4. DISTANCE and TIME

We've already seen that even travelling to other planets in our own solar system will involve years of travel for astronauts or potential colonists. Travelling at the fastest speed we have managed so far, which is 20 times the speed of a bullet, the following transit times would be required for travelling to the various planets in our solar system:

- Mars – 1 year 2 months
- Jupiter – 2 years 6 months
- Saturn – 4 years 1 month
- Uranus – 8 years 3 months
- Neptune – 12 years 10 months
- Outer edge of the solar system – 41 years

(Image: nasa.gov)

One of the major problems in trying to traverse such vast distances in space is designing a propulsion system that can accelerate a spacecraft to very high velocities and then to decelerate it at the end of its journey. To date, all our spacecraft have used chemical propellant rockets. They use chemical reactions to create a hot gas that is expanded rapidly to produce thrust. A significant limitation of this kind of propulsion method is that chemical reaction rocket engines are extremely inefficient; the ratio of thrust to the mass of the propellant is very low. In other words only a small percentage of the total energy of the propellant is converted into thrust; a lot of the propellant's energy is converted into heat and light rather than propulsion. This means that chemical reaction rocket engines need to carry enormous amounts of fuel. For example, the Apollo rockets used 1.9 million litres (500,000 gallons) of fuel just to leave Earth's atmosphere.

If spacecraft will need to accelerate for months or even years and then decelerate at the other end for the same amount of time, chemical fuelled propellant rockets are completely out of the question. For this reason, NASA and other organisations are now conducting research into more fuel-efficient propulsion systems. Many of these propulsion systems started as mere speculation by science fiction writers, but now, decades after their appearance in science fiction novels, various space agencies are starting to investigate the possibility of making them a reality. (As a matter of fact, science fiction has been the catalyst for many important human innovations, including submarines, helicopters, cell phones, robots, tasers, self-driving cars, 3D printers and atomic power. Each of these technologies were ridiculed when science fiction first proposed them!)

(Image: BBC.com)

Some of the possible future propulsion systems that are currently being investigated include the following:

ION DRIVE:

Accelerating ions and emitting a super-fast electron stream. Molecules of a stable gas such as xenon are ionised (given a positive or negative charge) and then accelerated by an electric field.

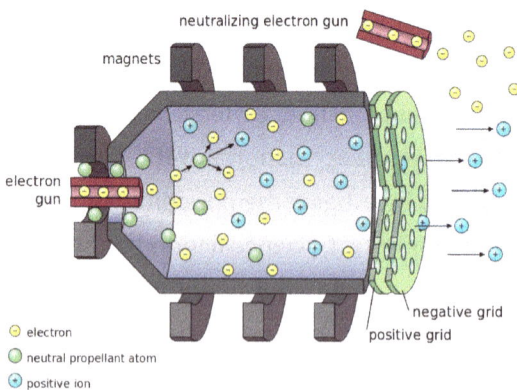

The charged particles are then fired out the back of the engine, thereby creating thrust. Several spacecraft have been constructed in recent years which use newly developed ion drives. These include "Deep Space 1" constructed by NASA (which flew past an asteroid and a comet) and "Dawn" (a NASA probe tat travelled to the asteroid belt, launched in 2007).

PLASMA DRIVE:

This drive involves generating thrust through extracting an ion current from a plasma source. This is similar to an ion drive, except instead of using a stable gas as fuel, magnets and electrical potentials are used to accelerate ions in plasma to generate thrust.

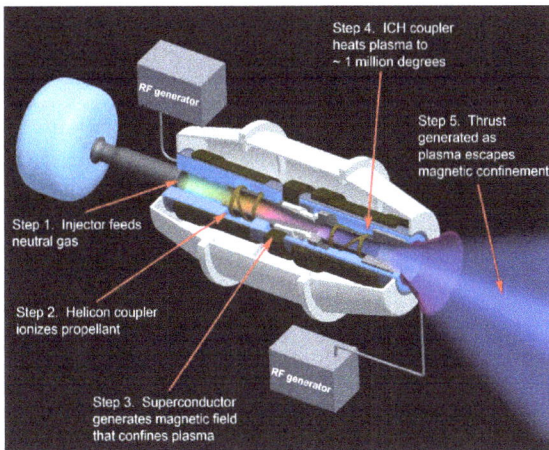

A plasma rocket has recently been constructed by the Ad Astra Rocket company in Texas and they claim it could enable a spacecraft to reach Mars in just 39 days.

.   .   .

SOLAR SAILS:

This is the concept of utilising solar radiation pressure exerted by sunlight on large reflective sails. Photons (particles of light) exert a tiny pressure. If a large enough solar sail could be constructed, it could capture enough photons to generate thrust.

The main downside of solar sails is that the further a spaceship is from a sun, the less photon pressure there will be and the less thrust the solar sails will be able to generate. The European Space Agency (ESA) is currently researching solar sails as a means of propulsion for spacecraft in the inner solar system.

FUSION DRIVE:

Using a controlled nuclear fusion reaction to produce thrust. A nuclear fusion reaction differs from a nuclear fission reaction. Fission involves splitting the heavy nucleus of an atom into two lighter nuclei, whereas fusion is the process of fusing two light nuclei together.

Both reactions produce vast amounts of energy which could be utilised as thrust, but it is believed that a fusion reaction has a better chance of being contained and controlled. Large, earth based fusion reactors are already a reality, and lighter fusion reactors for spacecraft are currently being developed. In my science fiction novel, "The Stars That Beckon" and its two sequels in the StarPath Trilogy, I incorporated a fusion drive as a propulsion drive in some of the spaceships.

ANTIMATTER DRIVE:

Using antimatter as a power source for producing thrust. Antimatter was once the realm of science fiction but scientists have now produced antimatter in laboratories.

There are three types of antimatter drives currently being investigated; a pure antimatter drive (that ejects charged pions as thrust), a thermal antimatter drive (that uses antimatter discharges to heat a chemical propellant) and antimatter power generation (where antimatter discharges are used to generate a huge electric power source for an electric space propulsion system).

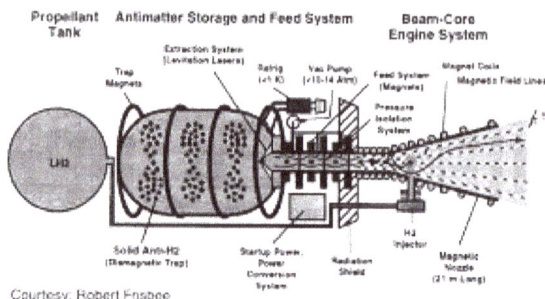

Courtesy: Robert Frisbee

In my science fiction novel, "The Stars That Beckon", and its two sequels in the StarPath Trilogy, I incorporated a pure antimatter drive as the main interstellar drive in some of the spaceships.

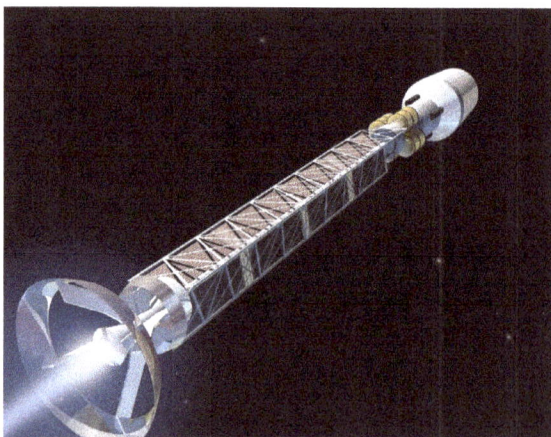

NASA concept drawing of an antimatter spaceship

BUSSARD RAMJET:

This concept involved gathering hydrogen from the interstellar medium via a ramjet scoop and compressing this until thermonuclear fusion occurs. This kind of propulsion system is named after the physicist, Robert W. Bussard, who first proposed it in 1960.

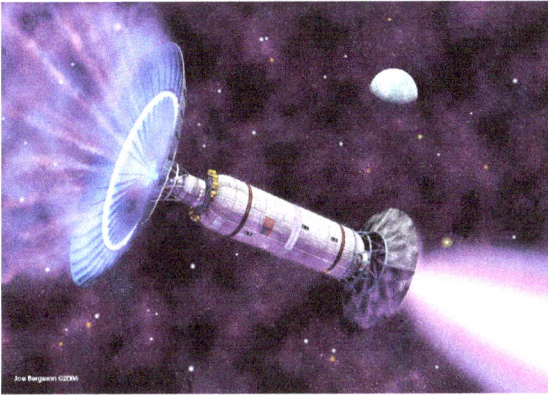

Bussard proposed using an enormous electromagnetic scoop (hundreds or even thousands of kilometres in diameter) to collect and compress hydrogen from interstellar space. The hydrogen is then forced into a constrictive magnetic field, compressing it until nuclear fusion occurs. The advantage of this propulsion system would be that the spacecraft gathers the fuel it needs as it flies through the cosmos. For such a propulsion system to work, however, the spacecraft would need to achieve an initial high velocity using some other preliminary propulsion system.

⁓

These are just a few of the 40 or more theoretical propulsion systems that are currently being discussed as possible solutions to the problem of accelerating and decelerating a vehicle of large mass over a period of many years. The goal is to design a propulsion system that produces more thrust using less fuel, so that a spacecraft can accelerate to much higher velocities more rapidly.

However, if such a propulsion system can be designed, whereby a spacecraft can achieve a velocity that is a significant percentage of the speed of light, another problem then comes into play; friction. When a spacecraft reaches more than 50% of the speed of light, friction from the particles of dust and gas in the universe becomes a problem.

5. FRICTION and MICRO-PARTICLES

Although space is a vacuum, it still contains atoms of hydrogen and helium as well as particles of dust and other left-over matter from the creation of stars and planets. The density of these particles is extremely low – about one atom every cubic centimetre – and at the kind of velocity mankind has so far achieved, these minute particles have no effect. But once we reach a significant percentage of the speed of light, a spacecraft will be travelling so fast that these microscopic particles will be creating friction on the hull which will heat it up.

As well as this, at such high velocities an occasional larger particle of matter (the kind that becomes a micro-meteor when it encounters a planet's atmosphere) would be a deadly threat, able to puncture the hull of the ship.

Thus, the future advent of powerful, more efficient propulsion systems will require the development of some form of shielding. Some scientists have suggested a physical shield made of super hardened metal. Others have suggested attaching a large asteroid or block of ice to the front of the spaceship, which would gradually be worn away throughout the journey.

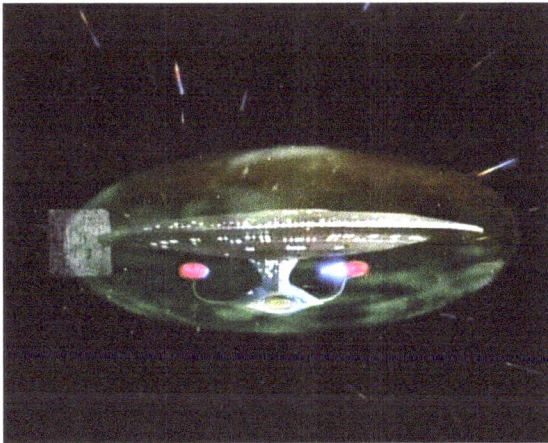

Still others have theorised the development of some form of energy shield. The challenge with all of these, however, is the deceleration phase of the ship's journey, when it will be flying backward and using its propulsion system as a brake to retard its velocity. During that phase a shield will be required at the rear of the vessel, yet this is where the propulsion engine is. Scientists are still scratching their heads about that issue.

∾

8

WHERE ARE ALL THE ALIENS?

There are many people who are convinced that the universe must be teeming with intelligent life. After all, if it happened once on Earth, it must have happened elsewhere, right? They argue that the universe is so big, there must be thousands or even millions of other intelligent species in existence.

(Image: Felix Huyn, glenn.zucman.com /

The SETI program (Search for Extra-Terrestrial Intelligence) is an international scientific endeavour to find extra-terrestrial life, and it has been searching the stars since 1984. In fact, individual scientists have been searching for aliens since the 1940s, long before the formation of SETI. So far, they have found nothing. No transmissions. No radiation signatures. No energy signatures. Zip.

A strong philosophical argument against the existence of aliens is the Fermi Paradox, first proposed by the astrophysicist Enrico Fermi in 1950. It refers to the apparent contradiction between the high probability estimates for the existence of extra-terrestrial life, and the ongoing complete lack of evidence for their existence. In essence, the paradox states, "Where are they all?" Fermi's argument goes as follows:

There are at least 200 billion stars in our Milky Way galaxy. Cosmologists estimate that about 90 percent of stars have at least one planet orbiting them. That's 180 billion. And, over the decades, we have observed that about one in five of the star systems that have planets include at least one roughly Earth-sized planet orbiting within the habitable zone, where liquid water can exist. That's 36 billion planets in the habitable zone.

Now, let's be extremely pessimistic and say that only one in 1,000 of those Earth-sized planets will ever develop life of any kind. That's 36 million planets with some form of life. And let's be equally pessimistic and suppose that, of those planets that do develop life, only one in 1,000 will go on to develop intelligent life that would eventually be capable of space flight. That's 36,000 planets with advanced, intelligent life. So, even using very conservative estimates, our galaxy should contain 36,000 planets with space-faring capabilities.

Enrico Fermi

Enrico Fermi, and his fellow astrophysicists in the 1950s were aware of similar estimates even back in the 1950s (although without some of the refinements that we are aware of today). One day, as they walked to lunch together discussing how the galaxy should be brimming with intelligent life, Enrico Fermi turned to his colleagues and said, "So where are they all?"

The paradox is made even more clear, when you consider that our own sun is supposedly a relatively young star, by astronomical standards. Cosmologists claim that up to 60 percent of stars in our galaxy are millions of years older than our sun (using evolutionary time-scales). We are relatively new kids on the block. So, according to scientists, 60 percent of planets which could have developed intelligent life are millions of years older than the Earth. That equates to 21,600 planets.

If these estimates are correct, there should be at least 20,000 intelligent species in our galaxy who achieved interstellar space flight millions of years ago. After all, it has only taken humanity thousands of years to venture into space. If these other species have had interstellar capabilities for thousands and even millions of years longer than we have, they would have spread out across the galaxy long ago. They should be everywhere!

So, where are they all? No one could answer Fermi's question in 1950, and they still can't today. We have been searching the galaxy for over 80 years now, and we've found nothing. With the most sophisticated scanning devices and telescopes on Earth and in orbit, we have found not a single trace of intelligent alien life. No transmissions. No spectral or radiation signatures of nuclear or any other form of power generation. Nothing. To date, we have identified thousands of exoplanets – planets orbiting other stars – and there are NONE that have all the life-promoting characteristics of Earth. They are all dead worlds, without any signs of life.

So, what am I saying?

I think we are alone in the universe. I think we are unique. I also think it shows that life doesn't just evolve by random natural processes, otherwise it would have happened all over the galaxy. I think it shows that we are a miracle; that we were created.

∾

9

THE UNIQUENESS OF EARTH

One of the exciting developments in recent cosmology has been the discovery of exoplanets – planets in other solar systems. It was long believed that other stars must have planets revolving around them, and the discovery of such planets over the last few decades has caused great excitement among cosmologists. It has also raised hopes of finding intelligent life somewhere out there! But are those hopes realistic?

Cosmologists agree that for ANY kind of life to be possible on a planet, a large number of conditions of habitability must be met, and all of these conditions must exist simultaneously on the same planet.

HABITABILITY ZONE

Firstly, a planet must orbit its star in what is called the habitable zone. This is the zone where the existence of liquid H_2O is possible. Water is essential to life. To date, no respectable exobiologist argues that life can develop without liquid H2O. But for liquid H2O to be present, a planet must orbit within a very narrow habitability zone of its star.

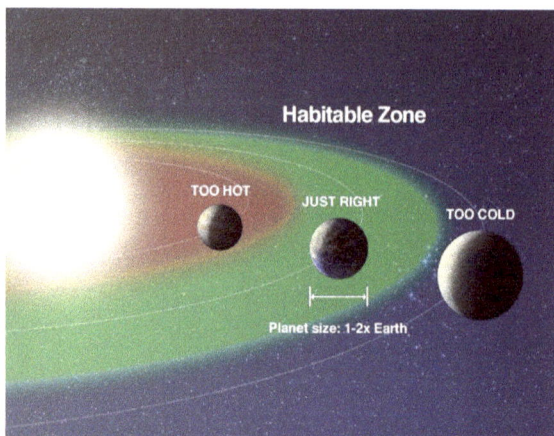

(Image: earth sky.org)

Too close to its sun, and liquid water will evaporate or boil away. Too far away, and water will freeze. The Earth lies perfectly within this habitability zone, or "Goldilocks zone" as it is sometimes called. If Earth was 5% closer or 5% further from the sun, biological life of any kind would not be possible.

TEMPERATURE

This goes hand in hand with the previous point, but it is worth mentioning separately. Cosmologists suggest that life is only possible within the temperature range, -35° C to +115° C. Below -35 C, chemical reactions become too slow to sustain life. Above +115 C, protein and carbohydrate molecules, as well as genetic material, start to break apart.

. . .

SIZE AND SPECTRAL CLASS OF STAR

There are many different types of stars in the universe, and the majority of them would be deadly to biological life because of the large amounts of dangerous and destructive radiation they emit.

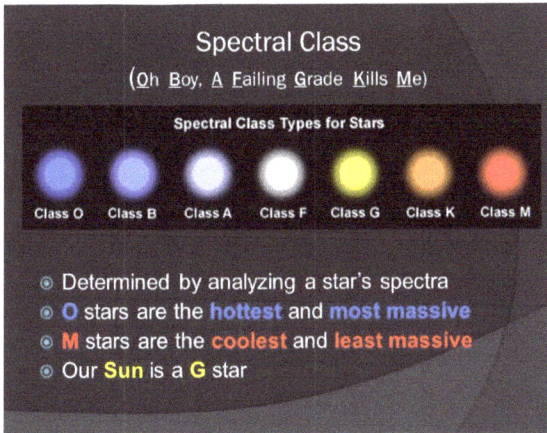

(Image: sideplayer.com)

The size and life-stage of the star is also crucial. Some stars are deadly in almost all their stages of development, with only a relatively brief time period when they would be safe for life to develop. Our own star, the Sun, is a perfect star class and stage for biological life. It is a G-class main sequence yellow dwarf. Cosmologists estimate that only about 5% to 10% of all stars in the universe would be suitable for biological life at any point in time.

LOW STELLAR VARIATION

The radiation from our star is relatively stable, but many stars have huge variations in radiation and luminosity, undergoing regular sudden and intense increases, making them life-prohibiting.

(Image: science.howstuffworks.com)

Some stars have an extremely unstable nuclear reaction taking place in their core which periodically produces deadly bursts of radiation that could strip away a planet's atmosphere.

THE RIGHT ATMOSPHERE

Small planets and moons have insufficient gravity to hold an atmosphere. Any gas molecules quickly disappear into space, leaving the planet devoid of any atmosphere. Very large planets with too much gravity will have an atmosphere that is too dense and heavy, also making it way too hot on the surface for biological life to develop. Our super giant planet, Jupiter, has an atmosphere that is thousands of times denser than Earth's at the point when the gas giant can be said to have any kind of surface. Even Venus, a planet of similar size to Earth, has an atmosphere that is 100 times denser than Earth's. As well as density, an atmosphere has to have the right chemical composition.

(Image: eapsweb.mit.edu)

Earth's atmosphere is 78% nitrogen, 21% oxygen, 0.9% argon and the rest comprised of trace elements. This is perfect for biological life. Venus, on the other hand, has an atmosphere comprised of carbon dioxide and other greenhouse gases, making it poisonous and also too hot for life. Scientists have discovered over 4,000 exoplanets to date, but none of them have an atmosphere that is life-permitting.

PLANETARY ORBIT

A planet needs to have an almost circular orbit, with very little elliptical deformation. If the orbit is too elliptical, it will either be too hot at one extremity or too cold at the other extremity for biological life to develop. Most of the planets in our solar system have extremely elliptical orbits, resulting in high variations in average temperatures throughout their year.

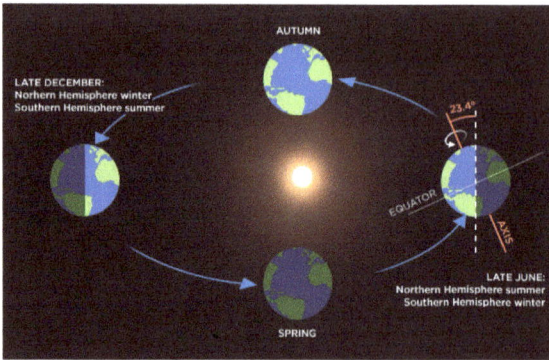

(Image: solarsystemscope.com)

Earth has an almost perfectly circular orbit, which makes it ideal for life.

PLANETARY ROTATION

A planet needs to spin on its axis if it is to be life-permitting. A planet that is tidally locked to its sun, with one side permanently facing its sun (like our moon), will be too hot on one side and too cold on the other for biological life to develop.

(Image: reddit.com)

Planets that have one side permanently facing their star are said to be tidally locked.

ROCKY

Most cosmologists agree that biological life probably can't develop in a gas planet. A solid surface is required to host life.

MOLTEN CORE

A molten core is essential for life. It serves three functions. Firstly, it helps to maintain a stable temperature on the surface of the planet.

(Image: Earth Science Stack Exchange)

Secondly, the movement of magma within the core creates a magnetic field around the planet, which protects it from deadly radiation from its star. Thirdly, it helps to circulate nutrients around the planet and recycle CO_2 from the atmosphere, provided there are moving tectonic plates.

. . .

MOVING TECTONIC PLATES

A mantle with tectonic plates that are in constant motion is considered to be essential for a life-permitting planet. The movement of these plates creates flows of lava from inside the planet to reach the surface, often deep under the oceans.

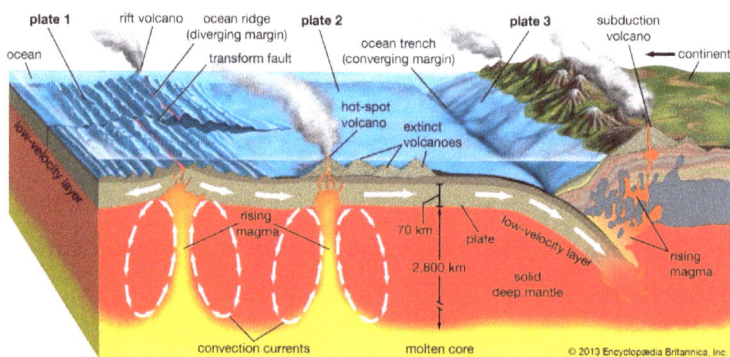

(Image: britannica.com)

These flows assist in circulating nutrients and ocean currents, as well as recycling CO_2 back out of the atmosphere by a number of complicated means. Without moving tectonic plates a planet with an atmosphere would spiral into green-house meltdown.

A LARGE MOON

Our large moon also helps to circulate nutrients and assist with CO_2 recycling through its tidal influence. Its gravitational influence upon our planet is quite significant, and it is becoming increasingly apparent that the moon plays a major role in stabilising and regulating our biosphere.

(Image: nasa.gov)

Interestingly, our moon is unusually large for a planet of our size, yet our large moon is a vital factor in the supporting of life on Earth.

WE WON THE COSMIC LOTTERY!

(Image: pixabay)

There are many other factors that are considered to be essential in order for a planet to be habitable and conducive to the development of biological life. Decades ago, scientists only spoke of a few essential factors, but more recently the list of essential factors for a life-permitting planet has grown to about 20. The fact that ALL 20 of these factors need to be present in the one planet for life to develop means that it is extremely improbable that a planet such as Earth would form by chance.

A recent published study by astrophysicist Erik Zackrisson from Uppsala University in Sweden calculated that the chances of a single planet having all 20 essential habitability characteristics is 1 in 700 quintillion planets. That's a 7 followed by 20 zeros. Zackrisson's findings are based upon a computer model that simulates the universe's development following an initial "big bang", using all the available data from our current understanding of physics and cosmology.

This is a big blow to the optimistic scientists who recently announced to the media that as many as 1 in 5 planets might be earth-like! One has to wonder what sort of mathematical gymnastics they engaged in to arrive at that figure, considering the fact that of the 4,099 exoplanets discovered at the time of writing (November 2019) *none* of them are Earth-like. That's correct – NONE.

The news media is currently abuzz with exaggerated claims about the latest and greatest exoplanet, Kepler-452b, the planet that is being touted as "Earth's twin". It has some similarities. It orbits a G-class main sequence star, similar to our Sun, and it is within the habitable zone of that star. But that is the limit of its similarity. What the scientists are not so keen to tell us is that the planet has a mass 5 times that of Earth and a diameter 60% larger than Earth, thereby making its gravity double our own.

On the next page is a picture of the planet published by the scientists so ready to extol it as our twin:

(Image: nasa.gov)

It looks impressive doesn't it? Blue ocean. Green land mass. It looks very inviting. However, the truth is that we have NO IDEA what the planet looks like. Kepler-452b is located 1,400 light years distant, which is much too far away for even our best telescopes to get the smallest glimpse of it. It is completely undetectable to us in the visible spectrum. The only way the planet was discovered was by examining perturbations in the light coming from that star and deducing that there must be a planet of that mass and size orbiting at a certain distance from the star. The drawing of the planet above is a complete figment of the imaginations of the scientists concerned!

Furthermore, the fact that a planet might be orbiting within the habitable zone of a planet is only 1 of 20 factors necessary for life to evolve. To date we have found dozens of planets much closer to our solar system which orbit within the habitable zone of their star, and none of them are even remotely Earth-like. Some are gas giants. Some are barren, rocky worlds without atmospheres. Some have spectral signatures indicating highly poisonous atmospheres. Some are tidally locked.

Despite the wishful thinking of scientists, many of whom are desperate to prove that life can and did evolve by chance, the evidence is proving otherwise. The more exoplanets we find, the more we realise how unique Earth is. Similarly, as we reach a deeper understanding of the many complex factors necessary for a planet to be habitable and life-permitting, we are inevitably drawn to one of two possible conclusions: Either we won the cosmic lottery, overcoming the most extraordinary statistical improbability that a planet like Earth should exist, or else our planet is not an accident – that it is the product of an intelligent designer who shaped it specifically for us.

∾

10

WHO MADE THE UNIVERSE?

T his brings us to the most important question of all. Where did the universe come from? Who made it?

THE EVIDENCE OF COSMOLOGY

(Image: nasa.gov)

Cosmology is the study of the universe, with its untold billions of galaxies and stars. For many people who look up in wonder into the night sky, the sheer scale and majesty of the cosmos has been regarded as a strong evidence for the existence of a Creator. Thousands of years ago, the Psalmist declared:

"The heavens declare the glory of God; the skies proclaim the work of his hands" (Psalm 19:1)

The response of the atheistic scientist to this cosmological "evidence" has been to simply claim that the universe has always been there. For centuries the theist has asked the atheist, *"Who made the universe?"*, to which the unbelieving scientist has replied, *"No one! It has simply always existed!"* Outspoken atheist, Bertrand Russell (1872 - 1970) once famously stated, "The universe is just there, and that's all!"[1]

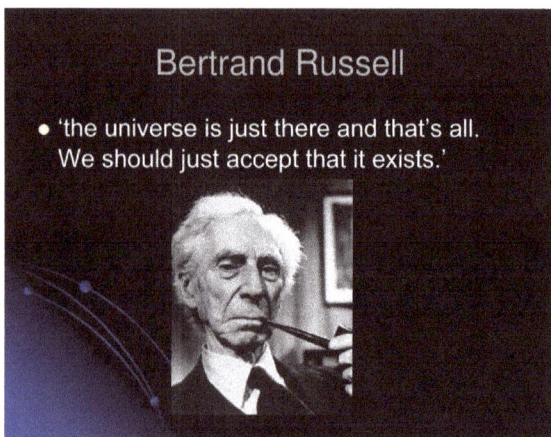

This belief in the eternal existence of the universe was the prevailing view of the scientific community from the time of

Aristotle (350 B.C.) until the beginning of the 20th century. This completely baseless presupposition was extremely convenient for atheists, because it effectively removed the necessity for believing in any sort of creator. A series of stunning discoveries in the modern era, however, has completely overturned this view of the universe and sent secular scientists scurrying back to their drawing boards.

After the publication of Albert Einstein's theory of general relativity in 1915, Dr. Willem De Sitter published extrapolations of Einstein's theory in 1917, predicting that, if Einstein's theory is correct, the universe should be expanding outwards and, conversely, must have had a finite beginning.[2] This idea of a universe that had a beginning was refuted vigorously by the scientific community at large, including Einstein himself, who wrote to Dr. Sitter, stating, "*This circumstance irritates me*" and in a second letter he stated, "*To admit such possibilities seems senseless.*"[3] The scientific community sided with Einstein, and largely ignored Dr. Sitter's theory.

In 1927, cosmologists Dr. Alexander Friedmann, Fr. Georges Lemaitre, Dr. Howard P. Robertson and Dr. Geoffrey Walker published more detailed extrapolations of Einstein's theory, (referred to as the Friedmann–Lemaître–Robertson–Walker metric)[4], confirming Dr. Sitter's findings that the universe must be expanding and lending further weight to the argument that it must have had a beginning.

In 1929, Dr. Edwin Hubble (1889 - 1953), confirmed the expansion of the universe by observing the Doppler redshift of the light from observable galaxies, indicating their recession from earth.[5] In other words, he confirmed from physical observation that the galaxies in the visible cosmos are expanding outwards.

Edwin Hubble

The discovering of cosmic microwave background radiation (CMBR) in 1964 by Drs. Arno Penzias and Robert Wilson[6] (which earned them a Nobel Prize in Physics in 1978) provided the scientific community at that time with what seemed conclusive evidence of residual microwave radiation left over from a "big bang" at the beginning of space-time. Thus, the Big Bang theory of the origin of the universe was formed. This concept that the universe had a beginning, represented a complete reversal of a belief in its *eternal* existence which had been passionately held for over 2,000 years. Dr. Stephen Hawking (1942 - 2018), in a lecture published on his website, commented on this reversal:

> *"All the evidence seems to indicate, that the universe has not existed forever, but that it had a beginning. This is probably the most remarkable discovery of modern cosmology."*[7]

BIG BANG IN TROUBLE

In recent decades, some serious doubts have been cast upon the veracity of a "big bang" as an explanation of the universe's creation. One of the major problems is explaining how such a big bang could account for the formation of galaxies and clusters of galaxies, which we observe throughout the universe. If a big bang created the universe, it should have resulted in a fairly uniform dispersal of matter throughout the universe, resulting in a sparsely spread scattering of matter.

(Image: hubblesite.org)

But this is not what we see. Instead, we find huge clusters of galaxies, densely populated by stars, separated by vast distances of empty space. No computer models of a supposed big bang can show how these clusters could possibly have formed. Dr. James Trefil, professor of physics at Mason University, Virginia, comments:

"There shouldn't be galaxies out there at all, and even if there are galaxies, they shouldn't be grouped together the way they are.' He later continues: 'The problem of explaining the existence of galaxies

has proved to be one of the thorniest in cosmology. By all rights, they just shouldn't be there, yet there they sit. It's hard to convey the depth of the frustration that this simple fact induces among scientists."[8]

Because of these and other observational anomalies that seem to contradict the Big Bang theory of the universe's origin, a growing number of scientists are calling for the theory's official demise. For example, in 1993, cosmologist Dr. Halton C. Arp, of the Mount Wilson Observatory, Pasadena, USA, wrote:

"In my opinion the observations speak a different language; they call for a different view of the universe. I believe that the big bang theory should be replaced, because it is no longer a valid theory."[9]

The expansion of the universe has also been challenged recently by some observational data which suggests alternate explanations for redshift in galaxies and for cosmic microwave background radiation (CMBR).[10]

These and other recent developments have led a growing number of respected scientists to reject the notion of a big bang. In 2004, an "Open Letter to The Scientific Community" by 33 leading scientists has been published on the internet,[11] and republished in *"New Scientist"*.[12] A subsequent article was published on www.rense.com, entitled *"Big Bang Theory Busted by 33 Top Scientists"*.[13]

Thus, modern cosmology is a long way from consensus on these issues. Even those who continue to hold to the big bang theory as the cause of the universe's beginning cannot adequately explain the observable complexity of the universe.

Something else had to have been at work, shaping the universe at the beginning of time.

THE UNIVERSE HAD A BEGINNING

Despite controversy and growing scientific scepticism regarding the big bang as an *explanation* for the universe's beginning, most scientists agree that the evidence of modern physics and cosmology points to the inexorable conclusion that the universe began *somehow*. The process of that beginning is unclear, but its fact is now almost universally accepted. This is further confirmed by several other very powerful theoretical arguments.

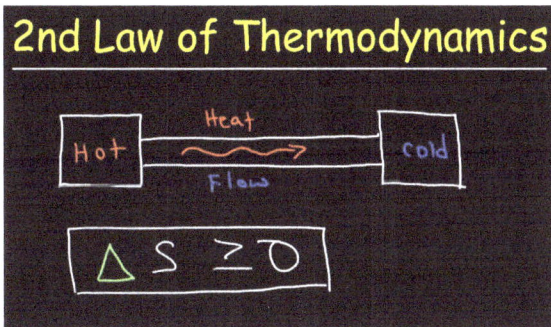

The Second Law of Thermodynamics

Newton's second law of thermodynamics indicates that in any closed system, the total amount of energy available diminishes over time. This is referred to as entropy, and, when applied to cosmology, it simply means that the universe is gradually winding down. In turn, this means that over time the universe is getting progressively colder. Our observation of the nature and function of stars confirms this. Because stars are finite bodies with finite mass and energy it is not possible for them to burn continuously for eternity. Given sufficient time, each star will eventually burn itself out of existence. This is a simple logical extension of Newton's theory. This, in itself, provides conclusive proof that the universe cannot have existed forever. If the universe has existed for eternity past, it would have burnt itself out by now!

The Philosophical and Mathematical Impossibility of an Eternal Universe

Al-Ghazali, a Persian philosopher in the Middle Ages, proposed this simple philosophical proof that the universe cannot have existed forever. He stated:

"If there were an infinite number of events in the past, we would never have arrived at the present"[14]

This simple, yet profound philosophical argument cannot be refuted without dismissing the laws of logic. Think about it. If time is a corridor and you started in the present and went back in time towards infinity past, you would never reach infinity past. You can't reach infinity past from the present. You will never get there! In the same way, you can't reach the present from infinity past. It's impossible! Al-Ghazali's theorem proves that the universe cannot have existed forever.

Similarly, Al-Ghazali also proposed that:

"An infinite number of things cannot exist."[15]

This is because infinity cannot logically exist in a physical universe. David Hilbert was a German Mathematician who came up with a brilliant way of explaining this concept. He asks us to imagine a hotel with an infinite number of rooms, each occupied by a guest. Therefore, the number of guests is infinity. Suppose one more guest arrives and wants a room. The hotel manager asks every guest to move to the next room number. The guest in room 1 moves to room 2. The guest in room 2 moves to room 3. The guest in room 5,000,000,000 moves to room 5,000,000,001 and so on. Now there is a vacant room in Room 1, and the new guest is accommodated. But the number of guests still numbers infinity.

Thus:

Infinity + 1 = Infinity

This works for the addition of any finite number. It also works for subtraction.

NO VACANCY BUT WE CAN MOVE SOME PEOPLE AROUND

But suppose an infinite number of new guests arrive, each requiring a room. No problem! Every existing guest is simply asked to move to a room number twice their existing number. Thus, Room 1 moves to room 2. Room 2 moves to room 4. Room 3 moves to room 6, and so on. In this way the pre-existing infinity of guests are now all in even numbered rooms, leaving an infinity of odd numbered rooms vacant. The infinity of *new* guests can now move into the odd numbered rooms.

But how many guests are there now? Still infinity! Thus:

$$\text{Infinity} + \text{infinity} = \text{infinity}$$

But suppose everyone in the even rooms now leave? An infinite number of guests were in the hotel. An infinite number of guests now leave (even numbers) but an infinite number of guests still remain (odd numbers). Thus:

$$\text{Infinity} - \text{infinity} = \text{infinity}$$

What on earth has all this got to do with the universe having a beginning? Quite simply, it shows how absurd infinity is as an actual number in a physical universe. Infinity exists only as an abstract concept; it cannot exist as an actual number in a physical universe. Therefore, there cannot have been an infinite number of past events and past days, hours or seconds. Therefore, the universe must have had a beginning!

The evidence of cosmology, philosophy and mathematics all point to a clear, logical conclusion: The universe has not existed forever. It had a beginning. As Dr. Stephen Hawking stated on his website:

> *"We have made tremendous progress in cosmology in the last hundred years ... which has shattered the old picture of an ever-existing and ever-lasting universe. ... This is a profound change in our picture of the universe and of reality itself."*

Dr Stephen Hawking

THE NECESSITY OF A SUPERNATURAL CAUSE

This brings us to a final, inescapable, astounding, logical conclusion; something *beyond* the physical universe had to have created it.

Most cosmologists are now reaching the astonishing conclusion that at some time in the distant past there was nothing at all. No stars, no planets, no asteroids, no gases, no chemicals, no elements, no physical matter at all. Nothing. And then a moment later, there was a universe! Stephen Hawking was correct in saying that this is "*a profound change in our picture of the universe and of reality itself.*" Because if the universe had a beginning, it raises the question: Who put it there? How can *nothing* become *something* unless *someone* beyond the universe created it? The cosmological evidence for the universe having a beginning cries out for a supernatural explanation, because only something *outside* of nature, something "supernatural", can possibly create nature. Nature cannot create itself out of nothing!

The Kalam Cosmological Argument

The Kalam cosmological argument deals with the issue of ultimate cause. It is a philosophical argument, originating in the Middle Ages, and championed in recent years by apologist and theologian, Dr. William Lane Craig. The original Kalam argument is as follows:

1. Whatever begins to exist has a cause.
2. The universe began to exist.
3. Therefore, the universe has a cause.

William Lane Craig added to the argument:

1. That cause must be timeless, immaterial, self-existent and powerful.
2. The most reasonable cause is God.

This is a logical argument that, in my opinion, is impossible to refute. The universe can't have created itself, therefore the cause must lie outside of the universe! Of course, there will always remain a core of obdurate atheists who refuse to concede even this clear chain of logic. For example, Dr. Quentin Smith, professor emeritus of philosophy at Western Michigan University, in a debate with William lane Craig, stated:

"The universe came from nothing, by nothing for nothing!"[16]

This, of course is a ludicrous proposition, and it illustrates that some atheists are prepared to believe in the impossible, rather than the supernatural.

A growing number of scientists, however, are concluding that there was something beyond nature that was fundamentally at work in the creation of the universe. In April 2016, Dr. Dan Reynolds wrote:

"All the observable evidence we have about the universe implies it had a beginning ... Logically, the universe did not and could not create itself. If the universe (nature) could/did not create itself and it had a beginning, then only something or someone outside of nature can account for the universe's existence. Genesis 1:1 offers a credible explanation: In the beginning God created the heaven and the earth."[17]

Dr Robert Jastrow, astronomer, physicist and founder of NASA's Goddard Institute of Space Studies, stated,

> *"Astronomers now find they have painted themselves into a corner because they have proven, by their own methods, that the world began abruptly in an act of creation to which you can trace the seeds of every star, every planet, every living thing in this cosmos and on the earth. And they have found that all this happened as a product of forces they cannot hope to discover. That there are what I or anyone would call supernatural forces at work is now, I think, a scientifically proven fact."* [18]

Dr Robert Jastrow

Similarly, Dr James Clerk Maxwell, physicist and mathematician, who is credited with formulating classical electromagnetic theory and whose contributions to science are considered to be of the same magnitude to those of Einstein and Newton, stated:

> "Science is incompetent to reason upon the creation of matter itself out of nothing. We have reached the utmost limit of our thinking faculties when we have admitted that because matter cannot be eternal and self-existent it must have been created."[19]

Commenting on the growing number of scientists who now concede that the universe must have had a supernatural cause, astrophysicist, Dr Hugh Ross, Director emeritus of Observations at Royal Astronomical Society, Vancouver, states;

> "Astronomers who do not draw theistic or deistic conclusions are becoming rare, and even the few dissenters hint that the tide is against them. Geoffrey Burbidge, of the University of California at San Diego, complains that his fellow astronomers are rushing off to join 'the First Church of Christ of the Big Bang.'"[20]

In other words, there is a growing tide of scientists at the top of their fields who, when confronted with the mounting cosmological evidence, are conceding that the only logical explanation for the origin of the universe, is that there must have been a supernatural cause. Nature cannot create itself; therefore the cause had to have been something outside of nature - a transcendent "supernatural" cause. This does not mean that all the scientists in the world are suddenly becoming Christians. There is a big step from believing in a supernatural cause of some kind, to believing in the God of the Bible. But for the first time in a long time, a growing chorus of voices within the scientific community has conceded the very real possibility of the existence of supernatural forces that lie beyond the realm of scientific study.

Dr Hugh Ross states;

"All the data accumulated in the twentieth and twenty-first centuries tell us that a transcendent Creator must exist. For all the matter, energy, nine space dimensions, and even time, each suddenly and simultaneously came into being from some source beyond itself. Likewise, it is valid to refer to the Creator as transcendent, for the act of causing these effects must take place outside or independent of them."[21]

Of course, this is not the impression that continues to be dished up to us by the media. The humanist movement has a stranglehold on the popular press. Scientists with aggressive atheistic agendas dominate the airwaves, and their well-financed documentaries continue to pump out the message that science has a natural explanation for everything. The impression is given that the new god of science has all the answers sewn up. But scientists who are at the cutting edge of their fields know that this is not so. Those who are studying these things in depth are coming face-to-face with undeniable evidence of a transcendent, supernatural reality that underpins our entire universe. The tide is turning, and it's not just a trickle. The previously quoted complaint of Dr. Geoffrey Burbidge, of the University of California at San Diego, that his fellow astronomers are *"rushing off to join the First Church of Christ of the Big Bang"*[22] indicates how widespread this new spiritual awareness is within the scientific community.

Without a doubt, the science of cosmology provides extremely convincing evidence for the existence of a supernatural creator-God. The extraordinary claim by Richard Dawkins, that *"there is not a tiny shred of evidence for the existence of any kind of god"*[23], must surely arise from a wilful determination to ignore the considerable cosmological evidence that has arisen in recent years, and portrays his intransigent unwillingness to even consider the existence of anything beyond the realms of science. One wonders how completely honest with himself he is being.

THE DESPERATE SEARCH FOR AN ALTERNATE EXPLANATION

As the evidence of cosmology increasingly points to a supernatural cause, die-hard atheists are becoming ever more desperate in their search for an alternate explanation for the origin of the universe that does not involve a supernatural Creator. These alternate explanations include:

- **Bubble universes:**[24]

This is also known as "eternal inflation theory".

This theory proposes that an endless series of new bubble universes, is expanding and breaking off from existing universes, like giant boils. But, surely, the question has to be asked: "Where did the first universes come from in the first place?"

(Image: sciencesource.com)

- **An oscillating or cyclical universe:**[25]

This is the idea that our universe is endlessly expanding then contracting back to a single point again, then exploding in a big bang again, on and on forever.

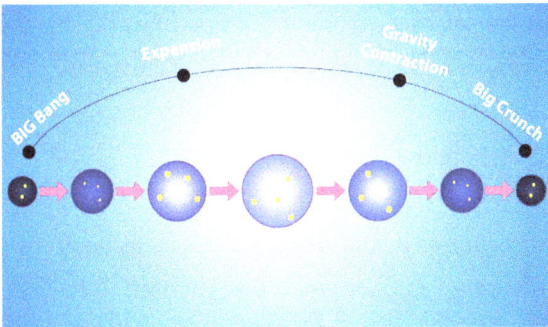

(Image: trihedron)

The problem with this theory is that both logic and the second law of thermodynamics have both shown that a physical universe cannot be eternal.

- **Baby universes:**[26]

One of Stephen Hawking's desperate postulations theorised that black holes are the umbilical cords that connect baby universes with the mother universes that gave birth to them.

(Image: thespaceacademy.org)

But, surely the question has to be asked: "Where did the mother universes come from in the first place?" This theory still does not answer the question of ultimate causality.

- **Aliens created our universe:**[27]

As ridiculous as it sounds, this theory has been proposed by Richard Dawkins and others. (Yes, Richard Dawkins actually proposed this in an interview that you can still watch on YouTube!)

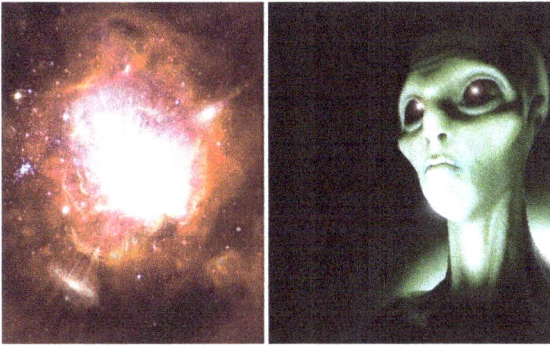

(Image: express.co.uk)

Do I really need to bother responding to this theory? I suppose I must. Please tell me, Dr Dawkins, who created the aliens??? This theory does not answer the question of ultimate causality.

- **Our future selves created the universe:**[28]

Oh dear! Now things are getting really silly, aren't they?

(Image: livescience.com)

Yes, some people have suggested that our future selves

eventually developed time travel and travelled back to the distant past and created the physical universe for our past selves to evolve into. (I'm not making this up! This has been suggested in scientific circles!)

- **Simulation theory:** [29]

(Image: inverse.com)

This theory proposes that our universe is simply a giant computer simulation designed by aliens or, once again, our super-advanced future selves, and we are all simply plugged into the simulation.

This theory was proposed by Dr. Nick Bostrom, professor of Physics at Oxford University in 2003.[30] ("The Matrix" is apparently true!)

- **Something from nothing:**

In another of Stephen Hawking's desperate attempts to avoid the existence of a creator-God, he postulated, "*Because there is a law such as gravity, the universe can and will create itself from nothing.*"[31]

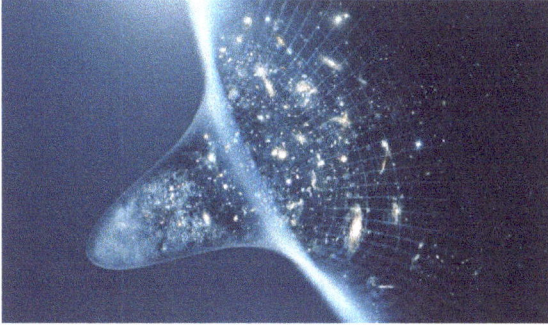

(Image: quantamagazine.org)

This theory effectively throws all the laws of science and the accepted laws of cause and effect, out the window! How can *nothing* create *something*?

- **We don't know:**

Those less inclined to scientific fairy stories, but still equally dismissive of the supernatural, simply state that we don't know how the universe was initially formed.

An article on the website, American Scientist, commented: *"The question of how matter came into existence in the formation of the universe still awaits a satisfactory answer."*[32] (At least this is honest!)

Christopher J. Isham, Britain's leading quantum cosmologist, and an astrophysicist at Imperial College of London, recently wrote:

> *"The idea that the Big Bang supports theism is greeted with obvious unease by atheist physicists. At times this has led to <u>wild scientific theories</u> being advanced with a tenacity which so exceeds their intrinsic worth that one can only suspect the operation of psychological forces lying very much deeper than the usual academic desire of a theorist to support his or her theory."*

Dr. Isham is absolutely correct. These wild and bizarre theories indicate a desperate determination to avoid belief in a supernatural creator, and a corresponding willingness to embrace anything, no matter how fanciful, in order to do so. Yet, as the cosmological evidence mounts up, it becomes increasingly clear that the universe could not have created itself. It had to have been created by something BEYOND nature – something supernatural.

Many people call that God.

～

PLEASE LEAVE A REVIEW

If you enjoyed this book, I would be extremely grateful if you would leave a review on Amazon. Reviews are hugely important for me as a self-published author. They impact Amazon's algorithms, helping the book to climb higher in Amazon's charts, thereby making it more visible to potential readers. Leaving a review is very simple and easy. Every single review really does help!

You can write a review by going to my Amazon author page and selecting the appropriate book:

https://www.amazon.com/-/e/B08295PT7V

(Sadly, Amazon only accepts reviews from people who have spent at least $50 on Amazon during the last 12 months).

SCIENCE FICTION NOVELS BY KEVIN J SIMINGTON

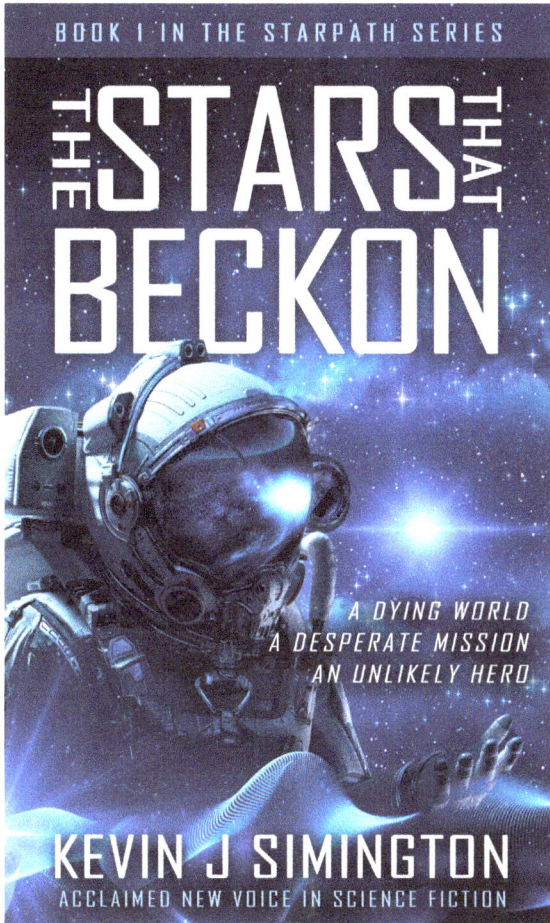

Grab a copy from any online book retailer!

(Available from December 2019)

BOOK 2 IN THE STARPATH SERIES

THE STARS THAT BEND TIME

THE SEARCH FOR A NEW HOME
JUST GOT COMPLICATED

KEVIN J SIMINGTON

ACCLAIMED NEW VOICE IN SCIENCE FICTION

Grab a copy from any online book retailer!

(Available from January 2019)

BOOK 3 IN THE STARPATH SERIES

A PATH THROUGH THE STARS

AN IMPOSSIBLE JOURNEY
A BATTLE FOR SURVIVAL
A SEARCH FOR HOME

KEVIN J SIMINGTON
ACCLAIMED NEW VOICE IN SCIENCE FICTION

Grab a copy Grab a copy from any online book retailer!

(Available from January 2019)

ABOUT THE AUTHOR

Kevin J Simington is a highly acclaimed writer whose fiction and non-fiction books have been lauded for their intelligence, wit and captivating writing style. He is a sought after keynote speaker who regularly speaks at conferences in the areas of philosophy, science and apologetics. He also writes for an international magazine. His "*StarPath*" science fiction trilogy has received glowing reviews.

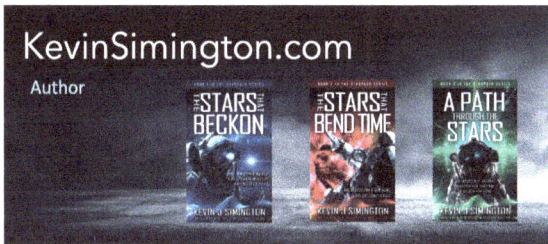

Follow Kevin on his website, kevinsimington.com, and on Facebook.

NOTES

10. Who Made The Universe?

1. Debate between Bertrand Russell and Fr. F. C. Copleston, broadcast on BBC radio, 1948, transcript: http://www.biblicalcatholic.com/apologetics/p20.htm
2. https://history.aip.org/exhibits/cosmology/ideas/expanding.htm
3. https://history.aip.org/exhibits/cosmology/ideas/expanding.htm
4. https://en.wikipedia.org/wiki/Friedmann%E2%80%93Lema%C3%AEtre%E2%80%93Robertson%E2%80%93Walker_metric
5. https://en.wikipedia.org/wiki/Hubble%27s_law
6. https://en.wikipedia.org/wiki/Cosmic_microwave_background
7. http://www.hawking.org.uk/the-beginning-of-time.html
8. J. Trefil, *The Dark Side of the Universe*. Charles Scribner's Sons, Macmillan Publishing Company, New York, USA, pp. 3, 55, 1988.
9. Halton C. Arp, quoted in E.P. Fischer (Ed.), *Neue Horizonte 92/93—Ein Forum der Naturwissenschaften—Piper-Verlag*, München, Germany, pp. 112–173, 1993
10. https://creation.com/expanding-universe-2
11. http://blog.lege.net/cosmology/cosmologystatement_org.pdf
12. Lerner, E., Bucking the big bang, *New Scientist* 182(2448)20, 22 May 2004
13. "Big Bang Theory Busted by 33 Top Scientists", www.rense.com, 27 May 2004.
14. Al-Ghazali, quoted in Greg dewar, "Advanced Philosophy and Ethics of Religion", Oxford University Press, Oxford. 2002, p.18.
15. Al-Ghazali, quoted in Greg dewar, "Advanced Philosophy and Ethics of Religion", Oxford University Press, Oxford. 2002, p.18.
16. Debate between William Lane Craig and Quentin Smith, https://www.reasonablefaith.org/media/debates/does-god-exist-the-craig-smith-debate-2003/
17. https://tasc creationscience.org/article/scientific-evidence-points-creator
18. https://en.wikipedia.org/wiki/Robert_Jastrow
19. James Clerk Maxwell; Perspectives on His Life and Work", Oxford University Press, 2014, p.274

20. Dr Hugh Ross, "The Creator and The Cosmos", Navpress, 2001, pp.108-112).

21. Dr Hugh Ross, "The Creator and The Cosmos", Navpress, 2001, pp.108-112).

22. Quoted by Hugh Ross, "The Creator and The Cosmos", Navpress, 2001, pp.108-112

23. Richard Dawkins interview, https://www.youtube.com/watch?v=of-8Q3HySjE&t=44m08s

24. Proposed by Paul Steinhardt and Alexander Vilenkin, in 1983. https://en.wikipedia.org/wiki/Eternal_inflation

25. https://en.wikipedia.org/wiki/Cyclic_model

26. Stephen Hawking, "Black Holes and Baby Universes", Random House Publishers, 1994.

27. Video Clip of interview between Ben Stein and Dr. Richard Dawkins, "Richard Dawkins Believes Extraterrestrials Created Man." https://www.youtube.com/watch?v=AiVoS78lNqM

28. https://www.abc.net.au/news/science/2018-09-02/block-universe-theory-time-past-present-future-travel/10178386

29. https://www.theguardian.com/technology/2016/oct/11/simulated-world-elon-musk-the-matrix

30. https://www.theguardian.com/technology/2016/oct/11/simulated-world-elon-musk-the-matrix

31. Stephen Hawking, cited in *Michael Holden (2010-09-02)*. "God did not create the universe, says Hawking". *Reuters*. Retrieved 2010-10-17

32. Article on Americanscientist.org, March 2017, no longer available.